No. 12 Ruston Crane Navvy of 1902. This was Ruston's first full-circle steam shovel, from which many hundreds of variations were developed with dipper capacities of from 3/4 cubic yard (0.57 cubic metre). Note here the chain hoist, soon to be replaced with cable. The driver (operator) has a nice warm place beside the boiler but needs the assistance of a cranesman.

Steam Shovels

Peter Manktelow

A Shire book

Contents

Cover: *A Ruston No. 300 steam shovel stripping the cover from ironstone workings near Kettering, Northamptonshire. The cover consists of layers of boulder clay, shales and sandstones which are upward of 50 feet (15.24 metres) thick. The machine is standing on the 9 feet (2.74 metres) of ironstone it has uncovered and is at the beginning of its excavation cycle. The racking (crowd) engine is exhausting from the head of the boom. This machine was purchased new in 1930 and worked until 1964, when this photograph was taken. A short time later it was cut up for scrap when mining in the area was deemed unprofitable.*

NOTE
The steam shovel was well established in the United States before its manufacture in Europe. The Americans by that time had already reached an agreed common terminology for the parts and functions of the machine. This has never been the case in Britain, with manufacturers voicing their own interpretation on nomenclature. For this reason no apologies are given for the use of American terms in this book. Alternatives are given only where deemed necessary to avoid confusion.

ACKNOWLEDGEMENTS
The author expresses his thanks for the help and support freely given by the many kind people in the excavating and earthmoving industries. He is especially indebted to T. W. Broughton and his colleagues at Ruston Bucyrus; Marshall Fayers, who imbued him with enthusiasm for old excavators; and Ian Hartland of the Vintage Excavator Trust, who gave him the necessary spur for this book.

British Library Cataloguing in Publication Data: Manktelow, Peter. Steam shovels. - (Shire album; 355) 1. Steam shovels - History I. Title 621.8'65'09. ISBN 0 7478 0483 4.

Published in 2004 by Shire Publications Ltd, Cromwell House, Church Street, Princes Risborough, Buckinghamshire HP27 9AA, UK. (Website: www.shirebooks.co.uk)

Printed in Great Britain by CIT Printing Services Ltd, Press Buildings, Merlins Bridge, Haverfordwest, Pembrokeshire SA61 1XF.

Ian Woollett's superb No. 6 Ruston steamer, painstakingly restored to its original condition. Weight of machine 17 tons (17.3 tonnes).

Introduction

The steam shovel, or steam navvy as it became known in Britain, has never been held in the same esteem in Britain as that accorded to most other forms of steam-driven machine. One has only to wade through the countless written works to discover the almost total disregard for the machine that could well be considered the precursor of all other machines. Without the excavating machinery in the open mines and quarries the economic extraction of the ores and rocks would be inconceivable and therefore no machines, steam or otherwise, would exist. Steam locomotives ran on railways built with the aid of steam shovels; docks, harbours and waterways where steam-driven vessels plied were dug out by steam shovels. In all civil engineering undertakings the work of the power excavator precedes almost all other operations. The steam shovel was designed originally to excavate earth, ore or gravel from its original location and load it into haulage units to be carried away (transported). If the material was too hard to be dug, such as rock, then blasting was necessary to break it down first; the shovel's single duty then was as a loading tool.

The steam shovel was the forerunner of a line of mechanical excavating machines introduced as a necessity and in good time for the great expansion of public utilities and mineral extraction that began in earnest in the nineteenth century.

The invention and successful application of the Otis steam excavator did not bring about an immediate revolution in the earthmoving business. In the first place the restrictive practices pursued through patent protection limited the manufacture of the machine and, when it became available, again it was by no means certain that the steam shovel would be readily accepted. Contractors were always slow to change their established methods and were reluctant to discard their picks, shovels and barrows, and horse-drawn scrapers and graders.

Once mankind had the raw materials and had developed the engineering skills to

Another view of the Ruston stripping shovel shown on the cover, dumping its load of 8 tons (8.2 tonnes). Machine weight: 350 tons (355.6 tonnes). Dipper capacity: 5 cubic yards (3.82 cubic metres). Boom length: 95 feet (28.96 metres). Dumping radius: 113 feet (34.44 metres).

make the machine tools needed for manufacture it was inevitable that he would seek more economical ways of moving earth. Although in Britain most of the great railways and canal networks were completed by the time excavating machinery came into general use, this was not so in many other parts of the world. Latecomers were quick to take advantage of the labour-saving benefits of machinery which allowed rapid progress in the development of irrigation schemes, transportation systems, waterways, dock and harbour facilities, and open mining and quarrying on an unprecedented scale.

The steam shovel was invented in the United States in 1835 out of the need to replace the large labour forces essential for the completion of railway contracts on time. Moving large volumes of earth with the primitive tools available in the early nineteenth century was both expensive and slow. After initial misgivings amongst the labourers, whose livelihoods were seen to be threatened by the newfangled 'American Devil', the original Otis shovel was finally accepted by the contractors. It was improved upon and developed without any fundamental change to its concept for almost sixty years.

For some time other types of excavators had been developed alongside the shovel, notably multi-bucket excavators in France and Germany. None of these machines, however, replaced the steam shovel, except in specialised usage. The machine that did eventually supersede the shovel in some duties was the dragline, but this was only after the cessation of steam power, when the dragline took on duties that were hitherto performed by the shovel. The British public, unlike the Americans, were probably never aware of steam shovels. Most of the basements and foundations of the large buildings and department stores in the USA were dug with the aid of steam shovels with the ubiquitous Mack 'Bulldog' trucks in attendance. An interested public looked on; the shovel-watchers were dubbed 'sidewalk superintendents'. There is

A Ruston Bucyrus B52 steam shovel in the 1960s at Shipton-on-Cherwell, Oxfordshire, digging limestone. This machine was one of the last steam shovels working in Britain and can now be seen in retirement in the Leicester Museum of Technology.

little photographic evidence of an equivalent situation in the United Kingdom. Is it that steam shovels in Britain were used mainly in the open pits and quarries, far from the public gaze? Every American knew what a steam shovel was; it had the same status as the steam roller in Britain and was the subject of many verses and narratives. In Britain today many diggers are called 'JCBs' by the unenlightened because of the proliferation of that type. When one reflects that the original steam shovels were part-revolving machines, JCBs are not so very different, except that they dig backwards.

The last shovel operated by steam was built around 1935 but these machines had a very long life and it is known that one was working in Britain until 1964.

Having limited our subject to steam-driven, single-bucket excavators, we have denied ourselves many other aspects of the fascinating and complex subject of earthmoving. What we see here is just the beginning.

A wooden side-tipping wagon of the kind favoured largely by British contractors for use with steam shovels. These wagons were known as the Manchester Ship Canal type because they were first used on that great undertaking. The wagon had a capacity of 4½ cubic yards (3.44 cubic metres) and was constructed for operation on rail track of the 4 foot 8½ inch (1435 mm) gauge. Photographed in a Kent chalk pit in 1968.

The very first steam shovel, the William S. Otis machine of the late 1830s. All motions worked from one single-cylinder, non-reversing engine.

The Otis Crane Excavator

The first efficient dry-land single-bucket excavator on record was the Crane Excavator designed in 1836 by the American William S. Otis, then aged twenty-three, a partner in the firm of Carmichael, Fairbanks & Otis, contractors from Canton, Massachusetts. Very few of these early machines were built before Otis died in 1839 at the early age of twenty-six. The manufacture of the machine was limited by patent protection for about twenty-five years. In 1864 the production of the Otis 'steam shovel', as it was now called, was continued by John Souther & Company of Boston and continually improved upon right up to about 1912.

The Otis steam shovel was built on what was termed the part-swing principle, that is, the crane, or excavating equipment consisting of boom and bucket (dipper) handle, was mounted on one end of a travelling carriage or car and given a limited swing or slew capability. The main carriage of the machine was made of wood with iron reinforcing, as was the crane. The boom was supported by a hollow cast-iron centre column or standard which, by means of a large rotating ring at the head of the column, enabled the boom and its excavating equipment to be swung through 180 degrees or thereabout.

On the upper deck of the machine was placed the main machinery, a vertical boiler arranged for wood burning and a non-reversing, vertical, single-cylinder engine of 9 inch bore by 12 inch stroke (230 mm by 305 mm). This engine supplied power to all the motions of the shovel, namely *hoist* for the dipper, *crowd* to thrust the dipper in and out, *swing* or slew, and *travel* via rail wheels of broad gauge. The hoist motion for the dipper, or scraper as Otis called it, was by chain from a grooved drum on a secondary shaft from the engine. The chain was led through the centre column via pulleys leading over the boom and down to the bail on the dipper, giving a 3x purchase. The dipper was lowered simply by gravity, checked by a brake and held when necessary by pawls.

The thrusting (crowd) motion of the dipper was the principal constituent of Otis's patent claim. The dipper and its handle pivoted halfway along the boom, resting on a drum over which a double chain was coiled, leading up and fastened to opposite ends of the handle, the handle being iron-clad to prevent wear. When the drum was revolved the dipper was wound in or out. The drum was driven from a countershaft by a 1:4 gearing. At one end of the countershaft was a loosely fitted driving bevel wheel with integral band-operated friction clutch member; at the other end was a pinion for the reduction gearing, with a hand-operated brake between the two components. In essence, the power for the crowd motion was supplied from the hoist chain pulley with integral bevel gear at the top of the iron column. An ancillary shaft leading from this bevel gear led down to the clutch drive on the boom. With the aid of two handles, one for the clutch and the other controlling the brake, the cranesman could operate the crowd mechanism.

The swing motion was accomplished by two chains leading from a chain drum on the deck to the large ring on the mast head. The chain drum was revolved by means of a large bevel wheel affixed to it and driven by two reversible bevel pinions. These pinions were loose on their shaft and could be engaged with the drum wheel by a sliding jaw clutch, thereby revolving the drum, and therefore the boom, in either direction. The drive was taken through a friction clutch to relieve any stresses involved. The two reversible bevels also drove the travel facility via spur gearing.

When in use, the stability of the shovel was enhanced by side-mounted braces with adjustable screw jacks. The machine was operated by two men: a runner, or operator, who attended to the main functions; and a cranesman, who controlled the thrusting or crowd motion with the clutch and brake handles. The cranesman was also responsible for tripping the latch to open the bucket door. The door was closed and latched by its own weight upon returning to the commencement of the next cut, a feature continued to this day.

Improvements

From the outset Otis's machine incorporated all the motions, or movements, that later became common throughout the era of rope-operated power shovels. Later

An Otis-type shovel at work in 1872. Note that the shaft drive to the crowd drum on the early Otis shovel is now replaced by chain drive, one of the improvements introduced by O. S. Chapman.

The Steam Navvy designed by Dunbar and built by Ruston, Proctor & Company Ltd from 1875 until it was superseded by their full-circle machines around 1912.

shovels bore little resemblance to this pioneer machine but, nevertheless, the functions of digging, crowding, swinging, dumping and returning for another cycle were identical.

Many improvements were made to Otis's design, all with a view to increasing strength and efficiency. These included all-steel construction, three sets of independent engines (for hoisting, swinging and crowding) and locomotive-type boilers with high steam pressures. Another noteworthy innovation was the idea of mounting the shovel machinery on to a standard railroad flat car so that the excavator could be coupled up to a train. This in turn eliminated non-standard rail gauges on the site. The limited-swing shovel henceforth became known as the railroad type or, in the United States at the time, the standard two-truck shovel.

In Britain, Ruston, Proctor & Company Ltd of Lincoln built their first Steam Navvy in 1875. This machine, like the Otis shovel, had limited swing but was of wrought iron and steel construction instead of the timber and wrought iron of the Otis machine. Other makers in Europe commenced building part-swing steam shovels, such as Chaplin, Andrew Barclay, Menck & Hambrock, to name a few.

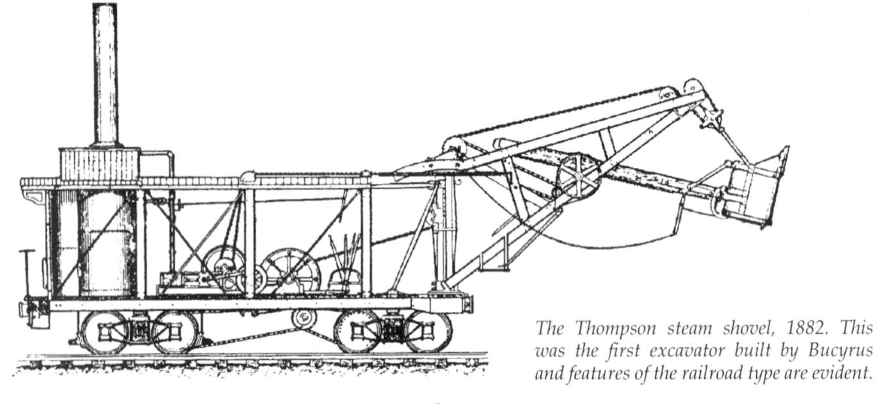

The Thompson steam shovel, 1882. This was the first excavator built by Bucyrus and features of the railroad type are evident.

The railroad shovel

This class of steam shovel was considered the ultimate development of the part-swing steam shovel, benefiting from around fifty years of improvements. The general construction and arrangement of machinery differed little between those made by Bucyrus, Marion, Osgood, Ruston and Vulcan, so the following description will serve them all.

The main framing, the car body, consisted of a heavy steel structure upon which rested the operating machinery and power source. The frame was mounted on two all-steel heavy-duty trucks (bogies), with three of the axles being driven by chains and gearing from the main, reversible engines on the deck of the shovel. The power equipment generally consisted of a locomotive-type boiler and reversible hoisting, swinging and thrusting (crowd) engines. The boiler worked at a pressure of 125 pounds and was provided with a duplex feed pump and injector.

Attached to the front end of the frame, or car, was the swing circle, upon which was mounted the excavator equipment, consisting of the boom, dipper handle, dipper and crowd engine. The boom was made in two sections, between which the dipper handle passed. The lower end of the boom was attached to the swinging circle pivoted to the front end of the platform. It was operated by cables, or chains, from the machinery. The boom assembly was supported by the A-frame and tie rods acting as tension members. The dipper handle was usually a single member of hardwood reinforced with steel plates. On its underside were fitted toothed racks that engaged with pinions on the shipper shaft.

The dipper was made of heavy steel plates reinforced where appropriate and provided with sharp teeth of manganese steel. The dipper door was hinged at its top and could be opened by hand; it carried a closing latch and closed by gravity as the

Principal features of a railroad shovel: 1, sills; 2, deck; 3, hoisting engine; 4, propelling chain; 5, screw jack; 6, trucks; 7, front end; 8, base casting; 9, swing circle; 10, crowd or thrusting engine; 11, shipper shaft; 12, handle racks; 13, dipper handle; 14, dipper door; 15, dipper; 16, dipper bail; 17, hoisting chain; 18, boom; 19, yoke block; 20, boom ties or guys; 21, head casting; 22, A-frame; 23, back leg; 24, control lever; 25, coal supply.

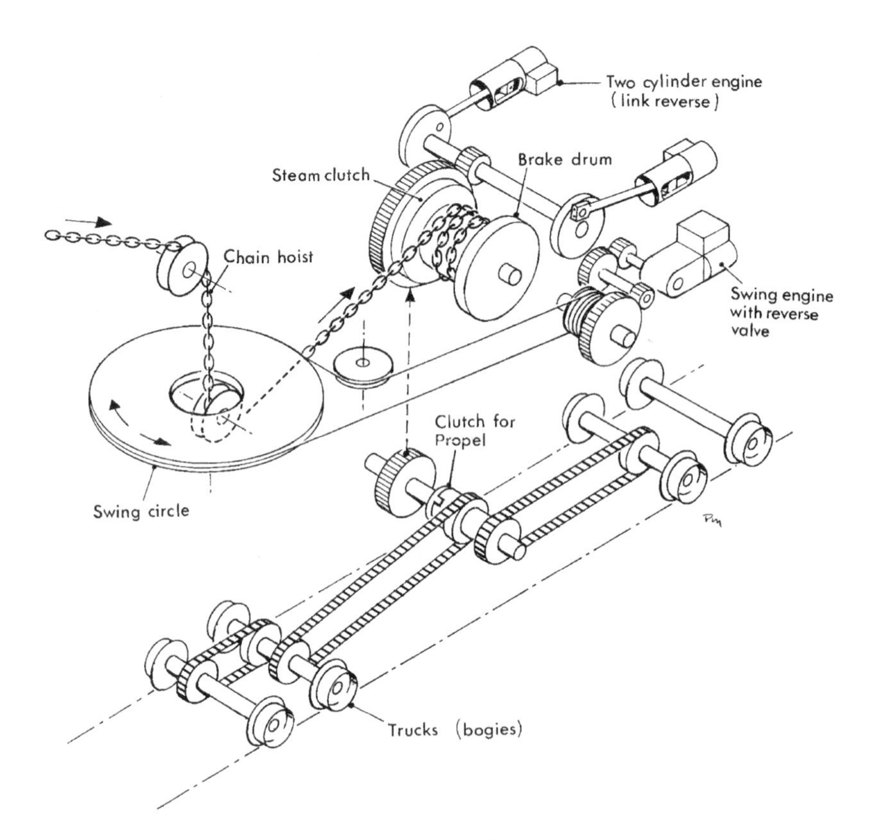

Schematic gearing diagram of a railroad-type steam shovel (Bucyrus). The crowd engine on the boom is not shown. The drawing is not to scale.

dipper was lowered for a cut. Railroad shovels were offered with dipper capacities ranging from 1$^1/_2$ cubic yards to 6 cubic yards (1.15–4.59 cubic metres) and with weights of from 43 tons to 120 tons (43.7–121.9 tonnes).

Below the A-frame members on each side of the machine were the jack braces or stabilisers. These were heavy steel castings carrying screw jacks at their outer ends. During the operation of the shovel the braces were placed at right angles to the car and the jacks screwed down on pads, preventing the machine from tipping and at the same time relieving the front trucks of digging loads.

The main hoisting engine on railroad shovels was of the double-cylinder horizontal type with Stephenson link reverse motion; as it did the heavy hoisting, it was the most powerful of the three engines. The engine shaft was fitted with discs with the cranks set at 90 degrees and a pinion that engaged with the large gear on the hoist drum shaft. The hoist drum was provided with housings at each end, one for the friction clutch band and the other for the brake band. This assembly was fitted with bronze bushings and ran loose on the hoist shaft. To operate the hoisting motion, the friction band attached to the large drive gear was tightened by a small steam ram so that it gripped the hoisting drum friction disc, making the drum revolve and so pull in the hoist chain.

The dipper was checked by the outside band brake on the hoist drum and was usually operated by a foot pedal at the operator's station. The hoist chain was taken to a pulley at the base of the swing circle and from there over pulleys on the boom to the bail block on the dipper. A three-part hoist was standard.

Plenty of steam! A Marion type 76 railroad shovel operated by the Stanton Mining Company of Scunthorpe, Lincolnshire. It was purchased in 1928 and worked in the Stanton mines for eighteen years. It worked an eight and a half hour day and performed its job at a rate of 1200 tons (1219 tonnes) per day. It was built in 1912 and was first delivered to a company in Nova Scotia.

The swing motion of the shovel was driven by a two-cylinder steam reverse engine with centre cranks set quartering and two eccentrics. The pinion on the crankshaft drove by means of intermediate gearing a gear on a drum grooved for wire rope. Two ropes led from the drum to the swing circle at the front of the machine. No brakes for the swing motion were necessary as it could be checked by shutting off steam and starting to reverse the engine. Because the crowd engine was displaced during the swing cycle the steam feed had to incorporate a swivel connection with glands, on the centre line of rotation. Also, as the hoist line was twisted, a chain was used in preference to cable.

The crowd or thrusting engine was usually a replica of the swing engine except that it had two pinions instead of one. These pinions engaged with large gears mounted on the shipper shaft, which in turn had pinions meshing with the dipper handle racks. The middle point of the shipper shaft carried a saddle block to maintain proper engagement between rack and pinion. The radius of the cut of the dipper was adjustable. The cranesman, who was situated on a platform at the foot of the boom, controlled the reverse valve of the engine and operated the dipper door.

The travelling motion of the whole machine was served by the main hoist engine by means of gearing from the hoist drum shaft. A jaw clutch engaged the propelling shaft, transmitting motion to the wheels via chains.

Compared with the Otis type of machine, the benefit of a separate, powerful engine for each of the three motions and the elimination of troublesome sliding jaw clutches meant smoother, more co-ordinated movements, causing less fatigue and therefore greater output.

The whole of the machinery deck was enclosed by housing of either wood or light steel construction and formed a protection from the elements for the boiler, water tanks, coal storage and at least one member of the crew.

A Bucyrus railroad shovel purchased in 1918 by the Oxfordshire Ironstone Company for work near Banbury. The largest machine of this type weighed around 130 tons (132 tonnes) and had a bail pull of around 49 tons (49.8 tonnes), which is impressive even by today's standards. (The bail pull of a steam shovel is the net pull on the dipper teeth with steam at blowing-off pressure in the boiler, the throttle wide open and the hoisting engines stalled.)

Operation

The railroad-type shovel was generally worked by a crew of seven men: an operator, engineer or runner (never a driver), a cranesman, a fireman, and four labourers (these are the terms used in the USA). The operator stood by a set of levers and brakes located conveniently to hand. The cranesman controlled the thrusting in and out (crowd) of the dipper, regulating the depth of cut and releasing the contents into haulage units or a spoil bank. (This two-man operation of the shovel was the same as on the Otis machine.)

The process of excavation began with the dipper handle in the near-vertical position with the dipper resting on the floor facing the cut. The operator moved the appropriate lever or levers setting the friction on the hoist, opened the throttle on the hoist engine and wound up the hoist chain pulling the dipper upwards. Simultaneously the cranesman operated the thrust engine, crowding the dipper into the bank sufficiently to ensure a full load. The operator declutched the hoist and set his brake, the dipper was retracted from the bank if necessary, and the operator then started the swing engine, placing the dipper over the dumping place. After the cranesman had pulled on the trip rope, releasing the contents of the dipper, the swing engine was then reversed and the dipper brought back to its position ready for the next cut. The action of lowering the dipper automatically closed the door, which latched under its own weight. After the entire face of the cut within reach of the dipper had been removed a new section of track was laid ahead of the machine and the whole unit was moved up 3 to 5 feet (92–152 cm). In order to do this, the jack screws had to be released and afterwards reset and the wheels blocked. The shovel was then ready for another cut.

As we have seen, the shovel excavator is an up-digging machine, that is, it digs in an arc away from itself and dumps its spoil to one side. As long as the limited-swing shovel was the predominant excavating tool, excavating patterns were dictated by the machine's limitations. The one serious drawback of the machine was its inability to

The Ruston No. 40 railroad shovel, which followed very closely the features of the Marion type 76 machine. The part-swing shovel never took a hold in Britain as it did in the USA, and Ruston built only a few of these machines before amalgamation with Bucyrus in 1930. Despite the advantages of the fully revolving types, the part-swing machine was nevertheless powerful, rugged, simple to understand and very accessible for maintenance purposes. Because it had less mass to rotate, its swing was very fast; no fully revolving shovel of comparable size could match its cycle time. Its poor mobility and limited development potential eventually led to its demise. It ceased manufacture in 1929.

dig in both directions. Once the whole length of the cut had been excavated within its reach, the machine had to be either bodily turned around for a pass in the opposite direction or run back to start another parallel pass. In each case the rails had to be shifted by a distance that depended on the reach of the shovel. Simply increasing the size of the excavator to enable it to dig more dirt on each pass was a limited option. Size limitations were a factor inherent in the design of all part-swing shovels.

Efforts were made to extend the usefulness and popularity of these machines by offering first broad traction wheel mounting and then crawler or caterpillar gear in order to reduce the work involved in track maintenance. Despite these modifications their days were numbered.

Whitaker & Sons of Leeds were the first manufacturers of the fully revolving shovel. On their first machine the crowd function was hand-operated and had limited stroke. The poor geometry of the boom hoist also affected its efficiency. This close-up makes an attractive family picture nevertheless.

A later Whitaker shovel, now with power crowd operated by a steam ram situated on the boom and worked via cranks. According to one story, these horses could uncouple the chain sling from the wagons without any prompting from the men. It was not clear whether they could put them back on!

The revolving shovel

The next significant improvement to the power shovel came with the introduction of the full-circle shovel which had 360 degrees swing or revolving facilities. Regardless of whether the undercarriage was arranged for rail, road wheels or crawler mounting, the steam shovel now had unlimited mobility. Excavating patterns could now become economically efficient and the machine could be designed accordingly. The first full-circle shovel was built by Whitaker & Sons of Leeds in 1884 and was adapted from a steam locomotive crane. Whitaker shovels did pioneer work on the Manchester Ship

A more efficient Whitaker shovel with much improved boom-hoist geometry and locomotive-type boiler for more steam. The crowd is still of limited stroke. Crane derivation is obvious here. We see that the workmen have time to pose for the camera.

A Wilson revolving steam shovel, 1887. The crowd is operated by a steam ram on the dipper handle.

Canal and also on railway contracts, with mixed success. Next came a machine built in 1887 by J. H. Wilson of Liverpool which for a time was successful both on dock contracts and in quarries. Ruston, Proctor & Company Ltd built their first full-circle machine in 1902, the forerunner of a long line of steam-powered excavators from this firm up to the 1930s.

In the United States the first full-circle machine was the Thew, followed by Marion, Bucyrus, Osgood, Erie, Menck & Hambrock, Wesserhutte and others. Once the advantages of the fully revolving shovel were realised, it was only a matter of time before the railroad shovel was relegated to history. Nevertheless, there was an air of caution amongst the large operators of the railroad shovel, especially as the standard two-truck shovel had been cleverly developed into a powerful and efficient machine.

Principal features of a revolving shovel: 1, crawler frame; 2, revolving frame; 3, crowd engine; 4, dipper handle racks; 5, dipper handle; 6, dipper; 7, bail; 8, hoist rope; 9, head sheaves; 10, boom stays; 11, yoke block; 12, shipper shaft; 13, water tank.

The Ruston No. 12 Crane Navvy of 1902. The machine had a bail pull of 12–14 tons (12.2–14.2 tonnes) and was fitted with a dipper of 2¼ cubic yards (1.72 cubic metres) capacity. Crowd (racking) was by steam cylinder on the boom, giving a stroke of 3 feet (914 mm). A two-man crew operated the machine.

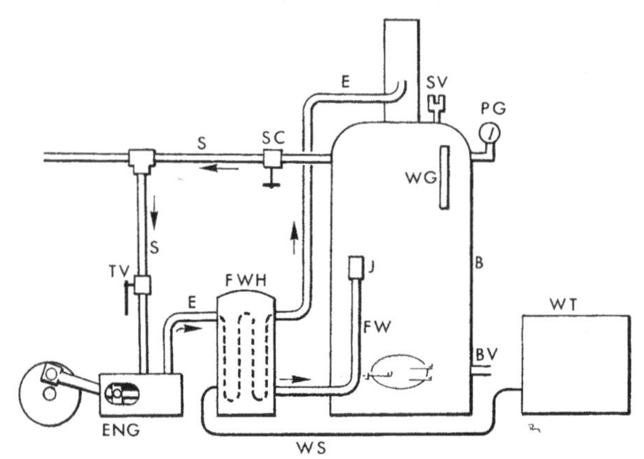

Steam distribution on a Ruston revolving shovel, showing how exhaust steam is utilised to heat feed-water into the boiler. B, boiler; BV, blow-off valve; E, exhaust; ENG, engine; FW, feed-water; FWH, feed-water heater; J, injector; PG, pressure gauge; S, steam supply; SC, stopcock; SV, safety valve; TV, throttle or regulator valve; WG, water gauge; WS, water supply; WT, water tank.

The mining companies in particular had considerable amounts of capital invested in good machines that still had long and profitable working lives and, though in the long run the expense of replacing machinery might be fully justified, such transitions would need great care and good judgement to implement. In Britain very few railroad shovels were introduced into the open pits and quarries because the fully revolving types were established from the start. In any case the open workings in Britain were never on the scale of the enormous copper, coal and iron workings elsewhere and these more modest requirements could be handled without too much inconvenience.

Initially the fully revolving shovel was limited to the smaller sizes, for the main reason of its compactness and speed of operation. It was particularly attractive to the smaller contractor, who used it to advantage in street excavations and in clay and gravel pits. It was relatively easy to move from one site to another in these small sizes, and competition amongst a growing number of manufacturers producing them gave rise to their popularity. There was the added attraction of the ease with which they could be converted to a crane or be fitted with other types of front-end equipment for a variety of digging applications. Another advantage was that they needed no additional support or jacks as the machinery on the revolving frame always acted as a counterweight against digging loads.

The very first full-circle stripping shovel, by J. H. Wilson of Liverpool, 1900: machine weight 70 tons (71.1 tonnes); dumping radius 60 feet (18.29 metres); dipper capacity $1\frac{1}{2}$ cubic yards (1.15 cubic metres). The crowd action was by steam ram between the dipper handles.

The long-range shovel

In spite of the many advantages offered by the fully revolving shovel, designers faced considerable technical difficulties. The development of larger sizes was slow, delaying the rapid changeover that had been thought a formality. In a fully revolving excavator the whole of the upper deck with machinery and excavating equipment has to revolve, and therefore there is a much greater mass to accelerate and decelerate on each digging cycle than with the railroad shovel. Unless more powerful engines are employed, the digging cycle will inevitably be slower, and the greater the power, the greater the stresses and strains. A balance has to be set between the expected higher

One solution for increasing the dumping range of a standard shovel was the use of a 'transporter' or 'tipple'; this was a structure carrying an inclined track upon which travelled a skip or trolley. This was a self-dumping device, the discharge point being controlled at will. In this scene a high-reach Ruston shovel, in the background, is loading the overburden into the skip, while a standard No. 20 Ruston machine is loading the uncovered ore.

Another view of the Ruston long-range stripping shovel uncovering an iron seam in Northamptonshire.

output of each cycle of the machine and the costs involved in achieving this end. That the large revolving shovel was successfully developed was due solely to the insistent demand from the mining companies for an economical method of uncovering or stripping to get at the iron ore, coal, phosphates or other material.

The early stripping shovels featured rail traction, as did the haulage systems, so it was advantageous for the shovels to remove large amounts of material before the tracks had to be moved. When excavating a channel or canal the large shovel could work at the bottom of the cut and load haulage units (trains or motor trucks) on either or both banks; track-shifting was then at a minimum. It was this saving that proved the worth of the large steam shovel beyond doubt. If large-scale earth removal was to be efficient, or extraction achieved without resorting to underground methods, there was no substitute for the large machine. The first so-called stripping shovel was designed and used in England, not the USA. Again J. H. Wilson was to the fore, building in 1900 two revolving shovels with extra long booms, or jibs. The next machine of this

Marion No. 350 8 yard (7.30 metre) stripping shovel in 1923 on iron-mining operations. At the time this was the world's largest and most powerful shovel. A total of forty-seven were built.

A Marion No. 350 operating machinery unit mounted on a one-piece cast-steel frame. The hoist clutch mechanism is evident on the large gear. Main engine: 14 inch bore by 16 inch stroke (356 by 406 mm). The cable drum is 48 inches in diameter (1219 mm); the brake drum is 14 inches wide by 84 inches in diameter (356 by 2134 mm).

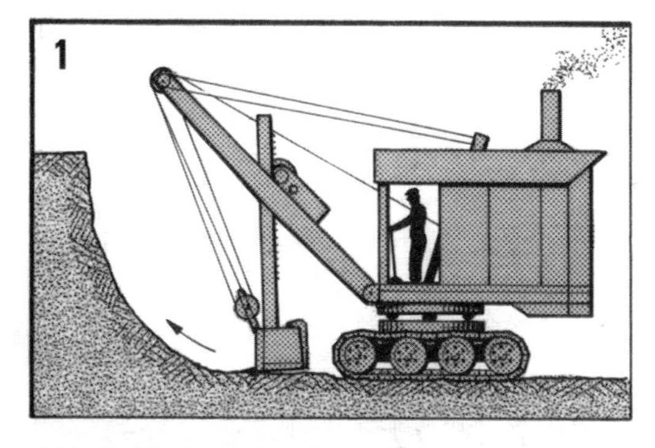

The operation of a full-circle shovel.

The operation commences with the dipper or bucket handle almost vertical and the dipper resting on the floor facing the cut. The hoist friction clutch is set, the throttle opened and the hoist drum is revolved, pulling up the dipper into the bank.

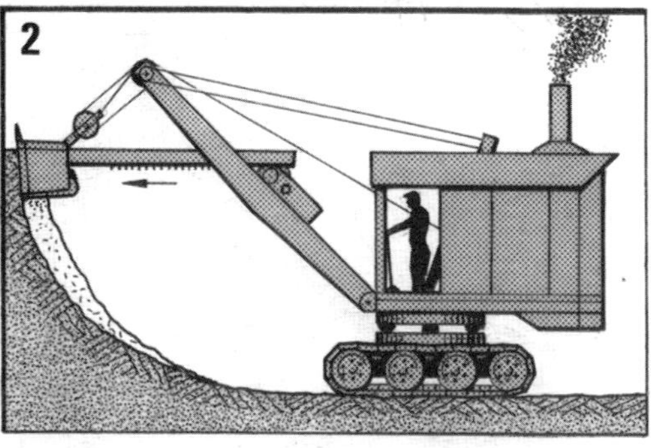

The crowd engine is operated, thrusting the dipper far enough forward to ensure filling. The upwards and forwards motions are performed smoothly and with co-ordination so as to avoid overloading the machine. The hoist brake is then applied to hold the loaded dipper. On reaching the top of the cut, the crowd engine is reversed if necessary, withdrawing the dipper from the cut.

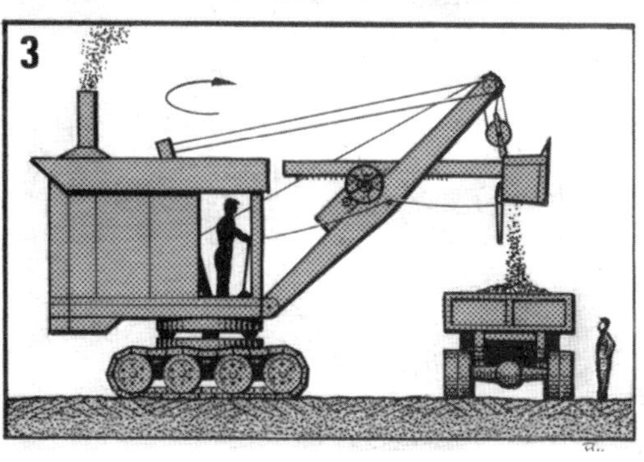

The swing motion brings the dipper over the dumping place. The trip rope is pulled, opening the dipper door and releasing its contents. The swing engine is then reversed, bringing the shovel back for another cycle, whilst at the same time the hoist brake is freed, allowing the dipper to fall back into position, closing and latching its door by its own weight.

The average time required for a complete cycle in ordinary conditions is 20–40 seconds.

type was designed and built by A. R. Grossmith in 1908. As far as can be ascertained, this machine incorporated the first ever rope crowd. Marion, in the United States, came later with a much larger machine, followed by Bucyrus and then Ruston, Proctor & Company Ltd. As regards the principle of shovelling, the dumping of spoil, plus most of the power equipment and machinery employed, the full-circle shovel functioned in much the same way as the part-swing machine. The prime difference lay in the greater mobility and the development potential of the full-circle machine. Gone now was chain for the hoist line, and one-man operation was to be the norm, with the 'driver' now facing the dipper at all times.

The range of machines offered settled into a fairly common pattern and only the sizes dictated differences in technical arrangements. Regardless of size, all machines had their boilers, hoist engines and swing and travel machinery on the revolving

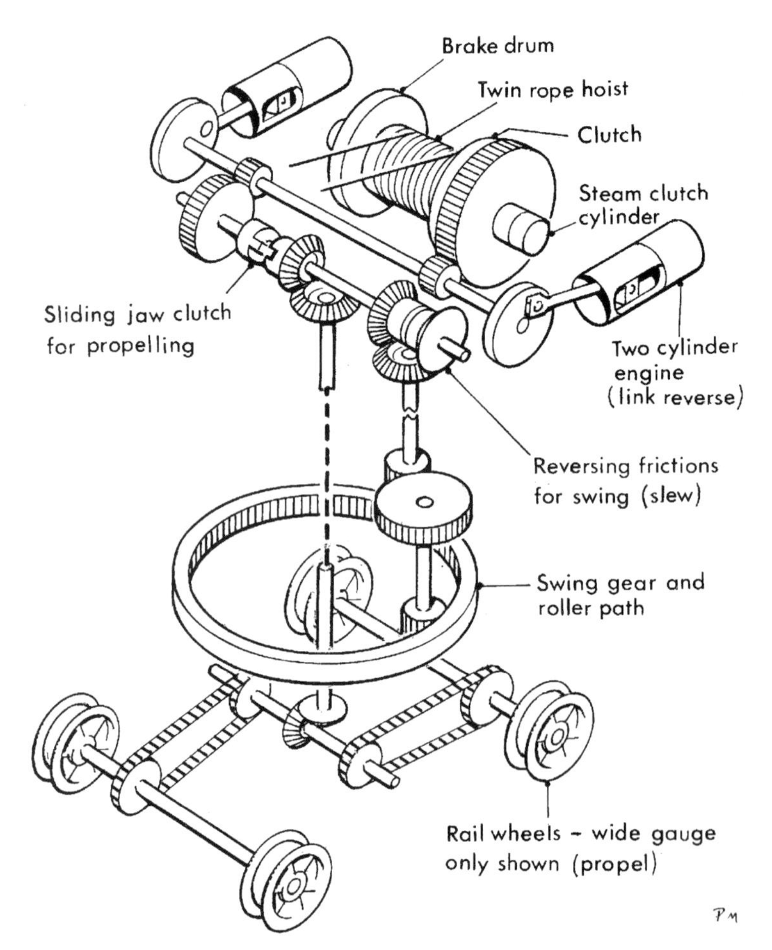

Schematic diagram of an early Ruston No. 20 gearing. Later versions had a separate engine for swing (rotate) motion, eliminating frictions. Optional crawler propel gear could be specified. (Drawing not to scale.)

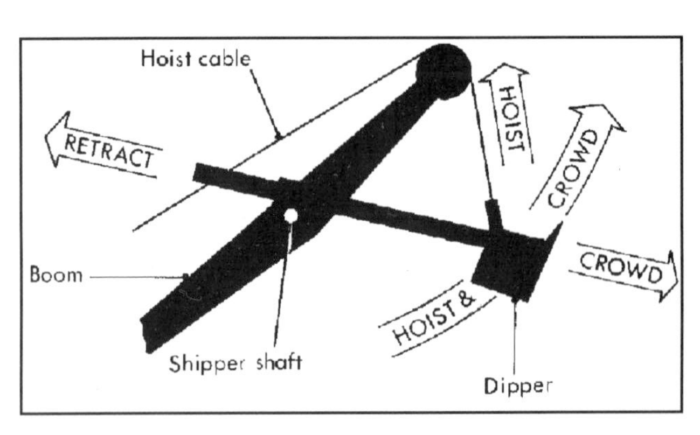

Shovel digging action.

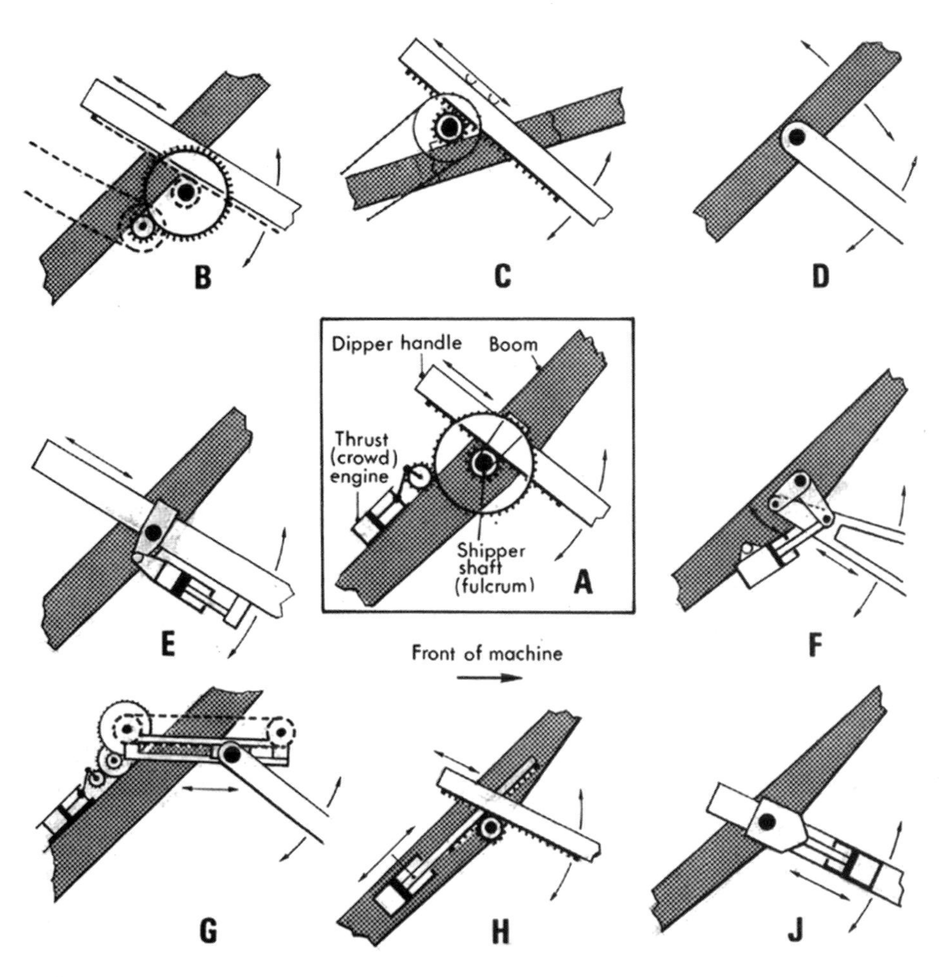

*Different forms of crowd arrangements on steam shovels. The crowd or thrusting motion regulated the amount the dipper thrust down whilst digging. It also thrust out or retracted the dipper to give the proper radius of dumping. Some of the many variations of this motion are shown here in diagrammatic form. **A**. Positive pinion and rack with reversing engine on boom. Dipper handles can be double either side of the boom or single within a split boom. First introduced by Barclay & Company, Scotland, on their steam navvy of 1877. **B**. Otis type powered by the main hoist line through a clutch take-off. The shipper shaft incorporated a chain drum for chain thrust on the handle. The drive to the chain drum is shown by chain, a modification by Chapman in about 1858. Otis's original model had shafts and bevels for this function. **C**. D. Carmichael and J. Osgood, 1846; Dunbar & Ruston navvy of 1875. A handwheel supplied the power to the shipper-shaft pinion on the boom, thence to a rack on the handle. **D**. Simple arrangement utilising a live or luffing boom with hinged handle. The crowd was not positive. First introduced on the Oswego Boom Machine patented by Sage & Alger in 1870 and built by the Vulcan Iron Works of New York. **E**. Steam-ram positive crowd by T. Dill, used on the Clement shovel in 1881. **F**. Whitaker's steam-ram lever arrangement, 1884. The stroke was limited to 2 feet (610 mm). **G**. The Thew shovel's crowd, 1895, featured a yoke member running on a horizontal track actuated by an endless cable or chain. Power was supplied by a separate reversing engine near the foot of the boom. **H**. Ruston Crane Navvy, 1902. It has a direct-acting steam cylinder on the boom with a rack driving a pinion on the centre of the shipper shaft and thence to rack pinions engaging racks on the dipper handles. **J**. The Wilson crowd, 1887, consisted of a direct-acting steam cylinder fitted between the dipper handles. On this machine the cylinder was thrust out with the handles, whereas on the Dill machine the piston was the moving member attached to the single handle.*

Types of clutches fitted to steam shovels. Clutches serve an important function in excavator gearing, transmitting power from one shaft to another at will. They can be classed as two types: positive clutches, which transfer the drive with no slip; and slipping or friction clutches, which allow a gradual and smooth take-up. A few examples are shown here in diagrammatic form.

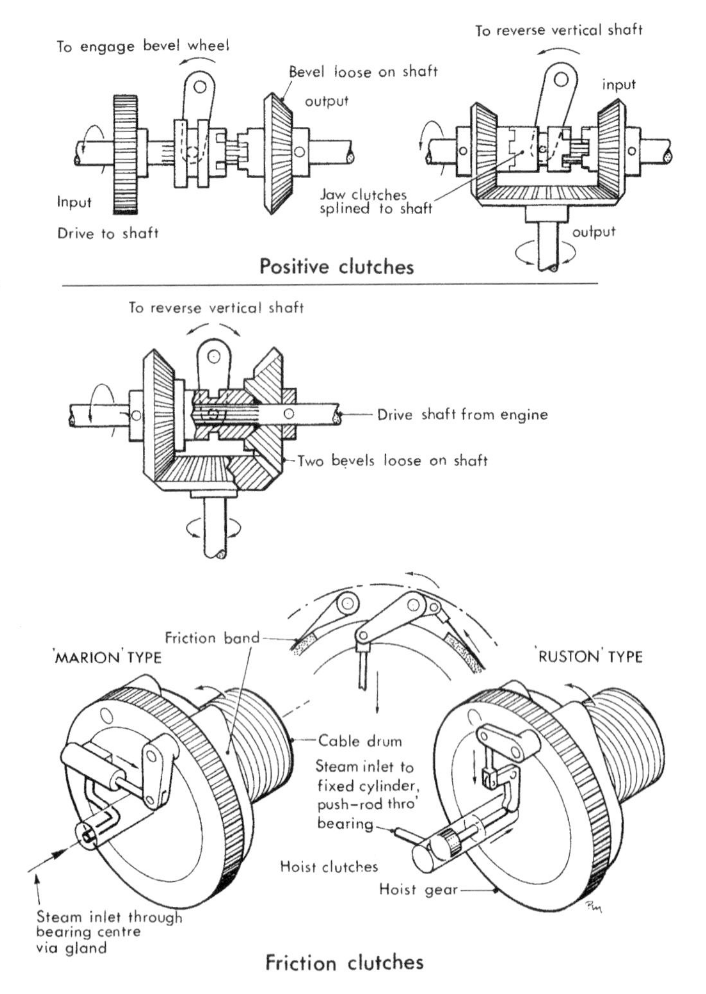

frame. On some makes the main hoist engine also supplied power for the propel (travel) motion whilst on others the propel motion was driven by the swing engine. In all cases the transference of power for travel was by means of jaw clutches. Some of the earlier machines in Britain had only one engine. There the functions of hoist, swing and travel were accomplished through clutches whilst the crowd was produced by a steam thrust cylinder mounted at the foot of the boom or on the dipper handle, giving linear motion. The ideal arrangement was with three engines, one for the hoist and travel, one for the swing, and another on the boom for the crowd.

All engines were of the two-cylinder reversing type, the main hoist engines having Stephenson link motion and throttle control. The hoist engines were of horizontal configuration, vertical, inclined or a compact V-type mounted on one side frame of the machinery. All cranks were quartered and valves were usually D-type although piston valves were latterly used in some cases, all worked from eccentrics. Fully enclosed, splash-lubricated engines were a feature on later machines and this included the crowd and swing engines. Various types of boiler were tried on steam shovels; the

Above: *A Ruston shovel of 1913, typical of the popular range numbered 15 to 20. Note the two-man operation of these earlier machines.*

Left: *The crowd engine on a Ruston navvy. A special reversing valve is situated above the steam chest. This enables the steam flow to be changed over from exhaust to inlet and vice-versa with the minimum of parts. Note the single eccentric for each cylinder. In Britain 'crowd' was called 'racking' for obvious reasons but it was not always the technically correct term. Rack and pinion was not always a feature of this motion.*

two basic types can be broadly classed as vertical and locomotive. The vertical boiler could be simple cross-tube or multitube. The locomotive multitubular boiler had the greater steaming capacity and was normally fitted to the larger machines.

Several types of mounting, or travelling, arrangements were available for these shovels although not necessarily on all makes, models or sizes. There were flanged wheels for single- or double-gauge rails, broad large-diameter traction wheels, and crawler or caterpillar gear in either two-track or four-track arrangement. The power for the propel, whether supplied by the main hoist engine or the swing engine on the shovel deck, was transferred to the base-frame drivers through the centre column of the machine via bevel gear, spur gears or chains. Rail-gear mounting needed no steering and was the simplest arrangement. At first crawlers were steered by jaw clutches operated from the ground but jaw clutches on later models were controlled from the cab by the operator (driver). With the exception of the power unit, all the structural features of the smaller classes of steam shovel should be evident to those familiar with later, diesel-powered machines.

The machinery for the hoist was much the same as that on the part-swing shovels except that wire rope now replaced the chain. Most of these revolving machines were

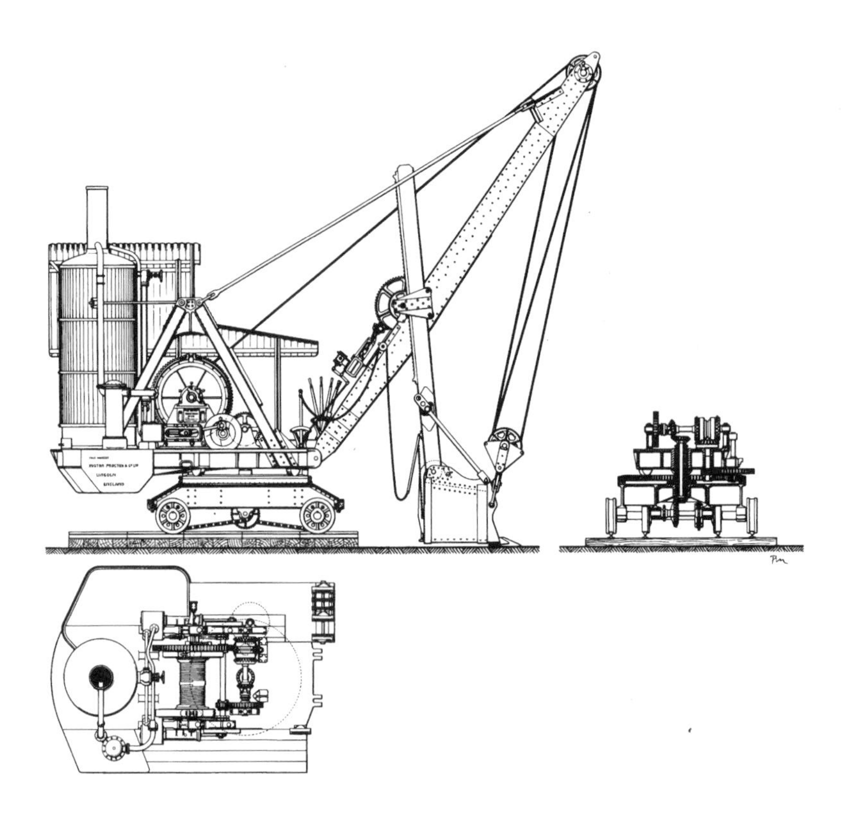

Above: *General arrangement drawing of a Ruston, Proctor & Company Crane Navvy of 1912. Ruston, Proctor classified their smaller machines on their ratings as cranes in terms of bail pull. For instance, if the machine, as a crane, could lift a load of 20 tons (20.3 tonnes) it was termed a No. 20 machine. This was only an approximate figure but served to size their shovels. The machine illustrated here is a No. 18 or an 18 ton navvy. In this drawing and the accompanying perspective drawing the main engine serves the hoist, rotate (slew) and travel functions. The crowd engine on the boom (jib) is a separate, reversible unit. Although the main engine has a link reversing gear, reversing of the slew facilities is done through separate, independent cone clutches. Thus slewing of the machine can be accomplished regardless of other functions in use. The main engine reverse facilities are needed for the hoist and travel functions.*

Right: *One of the last of the steamers. A Ruston Bucyrus RB 25 in a chalk quarry at Hessle, East Yorkshire. It was new in 1931. After lying idle for some years, this machine was dismantled and transported to Beamish, the North of England Open Air Museum, in County Durham, in the 1970s.*

A small steam shovel in its element, a situation where maximum efficiency can be expected. Nevertheless, the machine should be moved up – it is over-reaching.

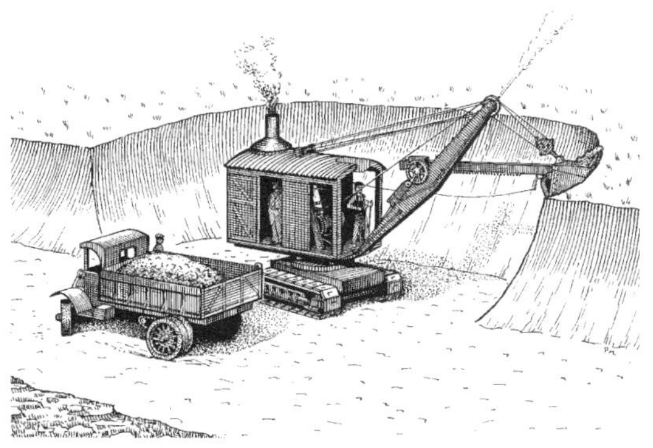

designed to be adaptable for dragline use; a special long boom was required plus a drag drum with attendant clutch and brake, a fairlead for the drag cable, and a boom hoist arrangement. Draglines are discussed in more detail in the following chapter.

The smaller sizes of revolving shovel came with dipper capacities of from $^{3}/_{4}$ cubic yard to $1^{1}/_{2}$ cubic yards (0.57–1.15 cubic metres), and with weights of from 20 tons to 50 tons (20.3–50.8 tonnes). The larger quarry-type machines came in sizes ranging from 2 cubic yards to 4 cubic yards (1.53–3.06 cubic metres) and weighed from 50 tons to 150 tons (50.8–152.4 tonnes).

The very large steam shovels of the time were designed principally for stripping the overburden from beds of coal, iron ore, copper and similar deposits. They had a working weight of up to 350 tons (355.6 tonnes), carried dippers of up to 8 cubic yards (6.12 cubic metres), used booms up to 90 feet (27.43 metres) long and dipper handles up to 60 feet (18.29 metres) in length. They usually rested on and were propelled by four four-wheel trucks of 3 feet (914 mm) gauge, one on each corner of the base frame. A suspension system was utilised allowing equalisation of loads on these trucks when on uneven surfaces.

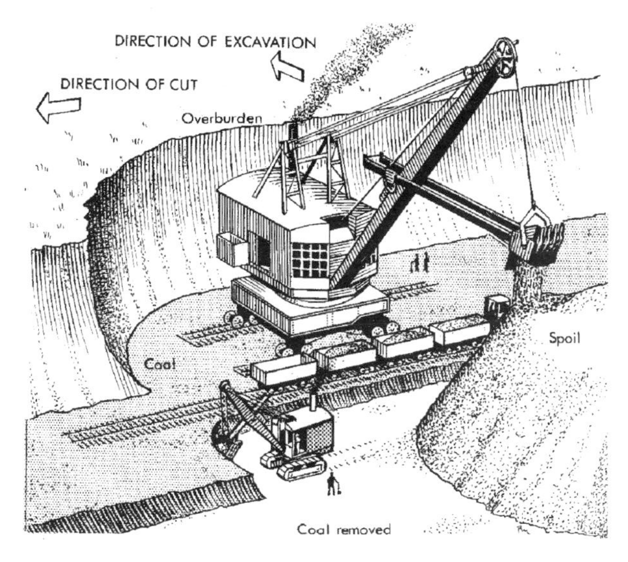

Typical open-pit coal-mining operation. The stripping shovel is sitting on the coal seam it has uncovered, whilst the small shovel is loading the coal into haulage cars (trucks). Note the long reach of the large machine, enabling the spoil to be dumped in the worked-out area of the pit.

An early dragline. This Bucyrus class 7 machine travels along the ground by means of skids and rollers. The machine moves ahead by pulling itself up to its bucket, which acts as an anchor when dug into the ground.

The dragline

The so-called dragline excavator was developed from the rather crude but effective land dredge or scraper excavator built for reclamation work in the middle eastern and southern United States. It was not until the modern form of scraper bucket (the dragline bucket) was patented by J. W. Page of Chicago that the attributes of the dragline machine were fully realised and appreciated. The principle of the dragline machine is that it drags its bucket towards itself and excavates almost entirely below the unmade surface upon which it stands. (The shovel, on the other hand, dug away from itself and excavated mostly above the surface it had prepared for itself.) The bucket of the dragline is suspended from a long boom and is lowered to the excavation before being dragged towards the machine. On being filled, the bucket is hoisted at the same time as the drag cable is being paid out, but only sufficiently to retain its contents until the machine is slewed or rotated for dumping. The bucket is suspended by the hoist line behind its centre of gravity and partly by an upper branch of the drag cable. Paying out the drag cable tips the bucket into a vertical position, thereby releasing its contents. Although the dragline excavator is more versatile than the shovel, it is neither as positive in its digging action nor as powerful, relying mainly on the weight of its bucket for penetration. It was its versatility, simplicity and range of action that were the reasons for its subsequent popularity.

The dragline boom takes no part in the digging function and is subject only to the weight of the loaded bucket plus the side strains of swinging the load. It can therefore

Operation of a dragline.

Normally the empty bucket is dropped from the point of the boom to the ground but an expert operator can throw the bucket some distance beyond this. This will increase the range of the machine but will slow down the cycle time.

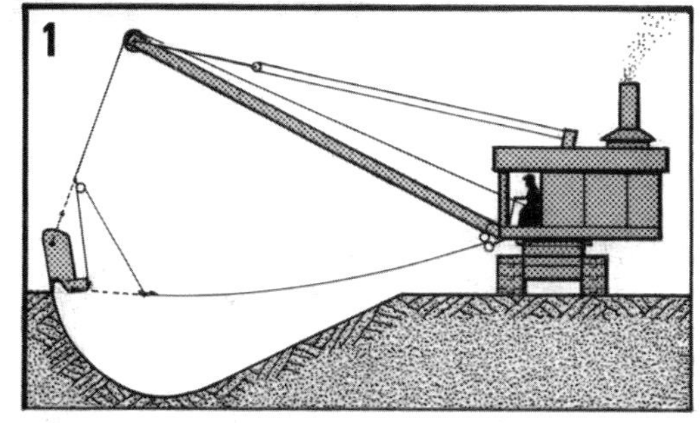

The drag cable is then hauled in, filling the bucket, whilst the hoist line is kept slack.

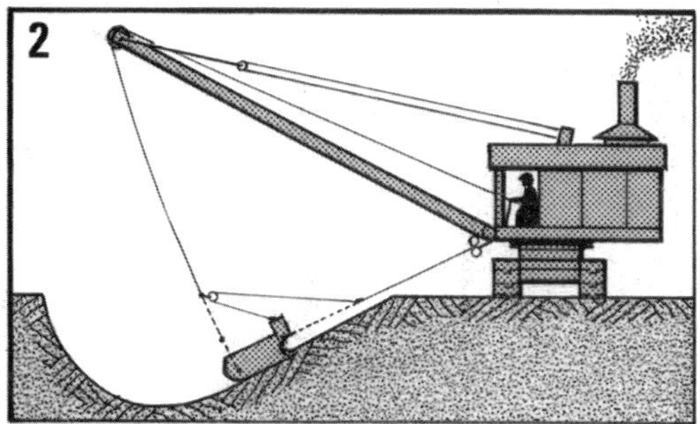

The hoist cable is wound in, lifting the bucket out of the excavation. The drag cable is slackened off to maintain the bucket in a horizontal position.

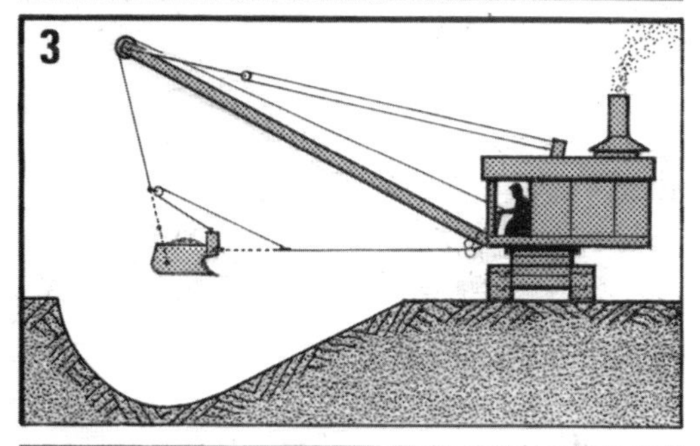

The machine is then revolved (slewed) to place the bucket over the dump. The drag cable is slackened off completely, allowing the bucket to hang on to the hoist line and dump its contents. The return swing is then initiated and at the same time the bucket is lowered for the next cycle.

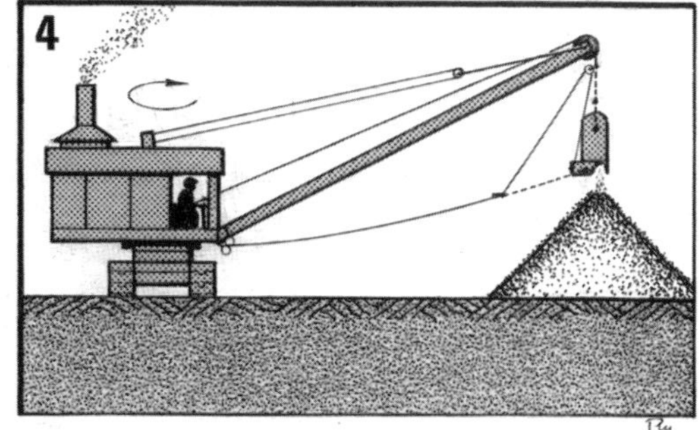

Sketch of a large dragline digging a canal and dumping the spoil to each side to form embankments.

be of relatively light construction and of great length. Weight for weight, the dragline carries a larger-capacity bucket than the shovel. Because of the nature of its dumping cycle it is not considered a loading tool and it relies on its great reach to dump its spoil without transportation needs.

The early draglines were rather primitive machines, built for one duty, but as the fully revolving excavators became established it was thought prudent to attune the basic shovel for dragline duties. Only when it became advantageous to do so was the dragline built as a specialised one-purpose machine in ever-increasing sizes. The first draglines relied on skids and rollers for their propelling function and used their buckets firmly dug into the ground for pulling themselves along. Later they were mounted on four-wheel trucks, one on each corner of the base frame like the shovels, or on crawlers. In 1913 Oscar Martinson, an engineer working for Page & Monigham of Chicago, invented the first practical walking device for the company's large draglines and this outmoded all previous means of moving these machines.

It will be appreciated that the fully revolving excavators have the advantage over previous part-swing machines in that their machinery, boilers and water tanks, affixed mostly behind the centre line of rotation, act as counterweight to their digging loads and bucket contents. They therefore have no need for further anti-tipping measures.

Early grab cranes: Priestman machines excavating the Manchester Ship Canal. Note the early type of bucket featuring the wheel and axle principle. Also note the wooden wagons, first used on this contract.

The grab or clamshell

There is one more form of single-bucket digging machine that, if not pre-empting the steam shovel, certainly was considered long before that machine's invention; this is the grab crane, or clamshell to give it its American name. The first efficient practical steam-driven grab crane is considered to be the creation of Priestman Brothers of Hull, who in the early 1870s were building and also exporting their steam cranes fitted with an early example of the double-scoop grab bucket of their design.

A Ruston machine with $1\frac{1}{2}$ cubic yard (1.15 cubic metre) bucket.

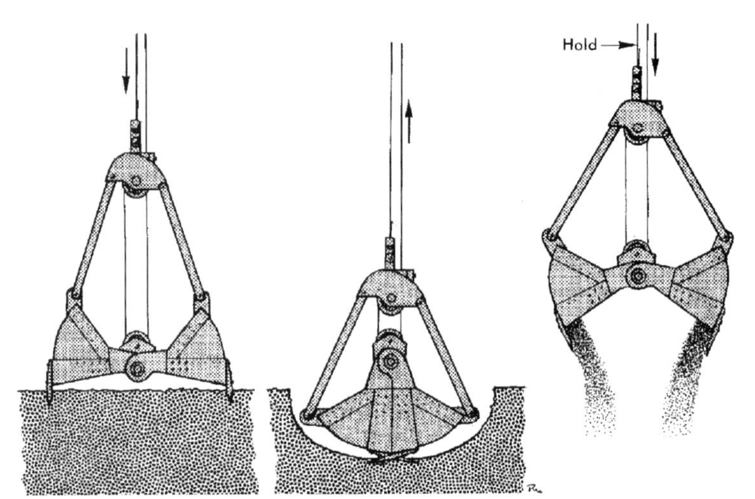

Grab or clamshell bucket operation. For efficient penetration of the material, excavating buckets must have sufficient weight and closing power to overcome any tendency to lift during the upward pull of the closing line. The heaviest concentration of weight of the bucket must be on the bottom sheave block, where it directly counteracts the lifting tendency when closing and creates a downward pull on the head of the bucket. This downward pull is transferred through the arms, pushing the outer lips of the shells into the material. The closing tackle on the early buckets featured what is termed the 'wheel and axle' principle. Later buckets used the 'block and tackle' or multi-rope principle, which allowed the reeving of as many as six parts of line for maximum closing power. Grab buckets have come in very many forms but for excavation duties it was customary to use either the two-scoop type with teeth or the 'orange peel' type. An anti-spin line is a requirement on all two-line buckets to obviate the tendency to spin, thereby fouling the lines.

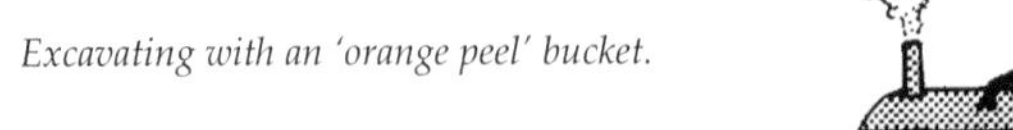
Excavating with an 'orange peel' bucket.

Grab cranes are now normally associated with the rehandling of bulk materials but they can, when fitted with an excavation bucket, be used for excavations of a kind that would preclude the use of other types of machine. They are primarily used for vertical excavation work below machine level in confined areas to a depth that is limited only by the amount of cable that can be wound on the drums. Grabs, or clamshells, are ideally suited for work such as digging foundations, footings, pier holes, trenches and cellars. They will dig equally well below water level; in modern dredging operations their work in this manner outweighs all their other uses as digging machines.

Two-line operation was established when grab buckets were introduced for excavation duties, earlier chain being superseded by cable. One line is the closing and hoisting line whilst the other is termed the holding line. This line holds the head of the empty bucket and is paid out to lower the open bucket on to the material. Closing power is then applied with the holding rope free. A crane with a long boom and two drums is a requisite for grab work. Only the closing drum needs to be powered; the holding drum can be friction-driven from the hoist drum shaft or wound in by a weight-driven line. To empty the contents of the grab, the holding line is braked whilst the closing line is run out, opening the two scoops of the bucket.

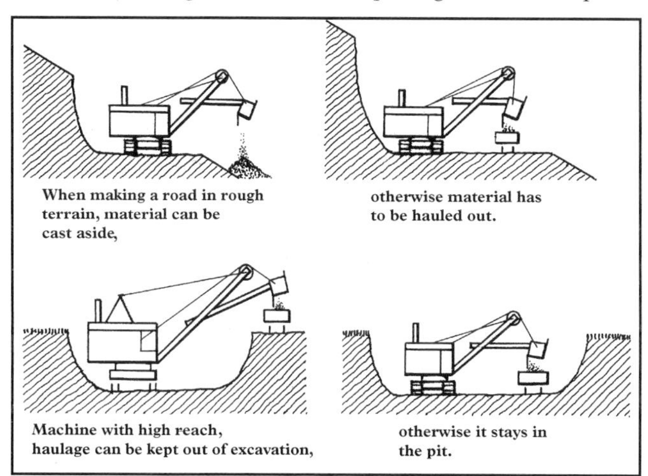

Typical excavating patterns.

Conclusion

Because of the difficulty in applying hand labour to underwater digging it was natural that the impetus for the development of excavating appliances began in this unfriendly environment. Steam power was first applied to an 'excavator' in 1796 in the form of a Boulton & Watt engine driving a 'spoon' dredger for use on Sunderland harbour. This predates the Otis machine by forty years, a span which would have been greater had suitable machinery been forthcoming at an earlier date. The rapid development of steam power in England resulted in more efficient dredgers and this meant the abandonment of the slow spoon dredger in favour of the multi-bucket dredger that had been used successfully for years in England and Holland, albeit animal- or man-powered. In the United States the simple single-bucket dredger or 'dipper dredger' was developed successfully alongside its dry-land counterpart, gaining the name of 'American dredger'.

Providing that they can be sited on firm ground, grab cranes and draglines can dig effectively under water. They may even be driven on to a barge or pontoon – the resulting unit being termed a dredger. The Otis-type machine lent itself in this fashion on more than one occasion. Specialised underwater excavators/dredgers, however, lie outside the scope of this book.

Steam power in excavators gradually gave way to more efficient and convenient sources of energy even though it was deemed by some to have been abandoned prematurely, before achieving its true potential and efficiency. This may be so, but so-called progress would eventually prevail over sentimentality.

Further reading

Barnes, W. *Excavating Machinery.* Ernest Benn, 1928.

Farrell, William E. *Digging by 'Stame'.* HCEA, USA, reprinted 1994.

Grimshaw, Peter. *The Amazing Story of Excavators* (two volumes). KHL Bookshop, 2002. (Plus other writings.)

Haddock, Keith. *Giant Earthmovers.* MBI Publishing Company, USA, 1998.

Marsh, Robert, Junior. *Steam Shovel Mining.* McGraw Hill, 1920.

Massey, George B. *The Engineering of Excavation.* John Wiley & Sons/Chapman & Hall, 1923.

Sheryn, Hinton J. *Illustrated History of Excavators.* Ian Allan, 1995.

Wislicky, Alfred. *History of Excavators and Dredgers.* Editions ATM, France, 1995.

Places to visit

Beamish, North of England Open Air Museum, Beamish, County Durham DH9 0RG. Telephone: 0191 370 4000. Website: www.beamish.org.uk (RB 25 steam shovel)

Leicester Museum of Technology, Abbey Pumping Station, Corporation Road, Abbey Lane, Leicester LE4 5PX. Telephone: 0116 299 5111. Website: www.leicestermuseums.ac.uk (RB 52 steam shovel)

Museum of Lincolnshire Life, Old Barracks, Burton Road, Lincoln LN1 3LY. Telephone: 01522 528448. Website: www.lincolnshire.gov.uk (Ruston & Proctor No. 12 steam shovel)

Threlkeld Mining Museum, Threlkeld Quarry, Threlkeld, Cumbria CA12 4TT. Telephone: 01768 779747. (Vintage excavators at work)

Early engraving of a Dunbar & Ruston Steam Navvy, 1875. The word 'navvy' was derived from the English term given to the manual labourer employed in cutting navigational canals. It is unfortunate that it was also applied to mechanical excavators, a usage that still prevails in Britain.